DOCUMENTED BRIEFING

RAND

Alternative Futures and Their Implications for Army Modernization

John Gordon IV, Brian Nichiporuk

Prepared for the United States Army

Arroyo Center

The research described in this report was sponsored by the United States Army under Contract No. DASW01-01-C-0003.

ISBN: 0-8330-3355-7

The RAND documented briefing series is a mechanism for timely, easy-to-read reporting of research that has been briefed to the client and possibly to other audiences. Although documented briefings have been formally reviewed, they are not expected to be comprehensive or definitive. In many cases, they represent interim work.

Published 2003 by RAND
1700 Main Street, P.O. Box 2138, Santa Monica, CA 90407-2138
1200 South Hayes Street, Arlington, VA 22202-5050
201 North Craig Street, Suite 202, Pittsburgh, PA 15213-1516
RAND URL: http://www.rand.org/
To order RAND documents or to obtain additional information, contact Distribution Services: Telephone: (310) 451-7002; Fax: (310) 451-6915; Email: order@rand.org

PREFACE

This documented briefing presents results from a RAND Arroyo Center research effort on how alternative futures could influence long-term Army modernization plans. The research was part of a larger project conducted for the Deputy Chief of Staff, G-3, U.S. Army, dealing with future force issues.

This documented briefing should be of interest to those in the Army who are concerned with long-range planning. The current Army Transformation Plan would take decades to implement. During that period, many changes could take place in the international environment that would influence the kind of Army capabilities that national leaders would need. Since there is no way to predict the future of 25 or 30 years from now, an examination of alternative futures could help guide Army long-range plans.

The research for this documented briefing was completed within the Arroyo Center's Strategy, Doctrine, and Resources Program. The Arroyo Center is a federally funded research and development center sponsored by the United States Army.

For more information on RAND Arroyo Center, contact the Director of Operations (telephone 310-393-0411, extension 6419; FAX 310-451-6952; e-mail Marcy_Agmon@rand.org), or visit the Arroyo Center's Web site at http://www.rand.org/ard/.

CONTENTS

SUMMARY

As the U.S. Army's current Transformation Plan makes clear, important decisions about service recapitalization will have to be made in this decade. This is because the Army's equipment stock is aging. Most of the major combat vehicles and helicopters now in the inventory are purchases from the 1980s and early 1990s. Those recapitalization decisions that must soon be made will have a long-lasting effect on the Army, as the systems that will be placed in development and/or production during this decade will be in the future Army for decades to come.

We believe that one approach to long-range planning that can help the Army chart its modernization course involves the use of alternative futures. The alternative futures approach has the distinct advantage of allowing the Army to hedge against the uncertainty represented by the international security environment of the far-off future. Planners can look for common equipment needs found across a range of plausible futures. This methodology also enables the service to clearly see a range of possible modernization requirements.

This alternative futures exercise is set in the 2020–2025 timeframe. We have consciously avoided assessments of the probability of each future, instead positing simply that each future meets the "not implausible" standard. Various "signposts" have been prepared to help determine which futures are becoming more plausible as time moves forward. An illustrative combat scenario was created for each future as well. Neither the signposts nor the scenarios are presented here, but they can be found in a companion RAND report.[1]

Six futures in total were developed to support this modernization study. The two best cases (from the perspective of U.S. national interests) are "U.S. Unipolarity" and "Democratic Peace." Our two medium-good futures are labeled "Major Competitor Rising" and "Competitive Multipolarity," and our one medium-bad world is called "Transnational

---

[1]Brian Nichiporuk, *Alternative Futures and Army Force Planning: Implications for the Objective Force Era*, Santa Monica, CA: RAND, MR-1676-A, 2003 (forthcoming).

Web." Finally, the worst case is a future entitled "Chaos-Anarchy." In each of the futures, a central mission for the Army was identified, a force size and structure was built, and necessary changes to the existing Army Transformation Plan suggested. The following paragraphs highlight these Army characteristics for each future.

The essential mission of the Army in a future characterized by U.S. Unipolarity would be to rapidly deploy for crisis-response missions in locations around the world. Opponents would be either regional powers of the same ilk as today's rogue states or yet-smaller actors. The overall active Army was sized to have a personnel end strength approximately equal to today's in the U.S. Unipolarity future. The Objective Force component of the force was modestly increased over current plans, while the Army XXI component was slightly reduced. Investment in deep strike and C4ISR saw small increases.

Democratic Peace is a very placid future and thus required an Army about one-third smaller than that of the 2000 era. Here, we made significant reductions in traditional Army XXI forces and truncated the Objective Force organizations currently being explored by the Army. On the other hand, the interim medium-weight units that are now being created would be well suited for this low-threat future.

Major Competitor Rising would require the Army to deter, and perhaps even defeat, a peer-level opponent. Major Competitor Rising would demand a 10 percent increase in Army end strength and major increases in the Objective Force, TMD, attack aviation, deep strike, and C4ISR. Medium-weight interim units would decrease from the total currently envisioned in the Army's Transformation Plan.

The Competitive Multipolarity future would call for the largest of our six future armies—50 percent greater in terms of end strength than the 2000 Army. In this future the Army would have to be prepared for deployments to confront hostile actions by two different coalitions that are opposed to the United States. In view of the need for rapid deployment capability and the requirement to oppose capable heavy units in this future, we envisioned a significant increase over the number of currently planned Objective Force units as well as increases in TMD, aviation, deep strike, and C4ISR capabilities.

In the Transnational Web future, the threats to American interests are fundamentally different from those we see today. Major portions of the U.S. armed forces have been reoriented to respond to the new challenges posed by hostile transnational actors such as organized crime syndicates and activist networks within certain ethnic diasporas. The Army in this future would be 40 percent smaller than today's. Here we determined that there would be a reduced need for Objective Force and Army XXI forces. SOF capabilities, however, would require a very significant increase over today's levels.

Finally, Chaos-Anarchy represents a future replete with many failed states and increased warlordism in much of the developing world. This future would require an Army end strength 10 percent smaller than today's. In this case, we made substantial increases in SOF, counter-WMD, and CSS capabilities, while C4ISR would merit a modest increase. Both the Objective and Interim Forces are increased modestly in Chaos-Anarchy as well. TMD and deep strike investments could be reduced for this future, since U.S. interventions would mainly encounter low-tech military forces.

Looking across the six futures, we found we could make a handful of general insights about the demands upon Army modernization between now and 2025. It did appear that medium forces (both Objective and Interim) are appropriate in a majority of possible futures. On the other hand, continued investment in Army XXI capabilities seemed necessary in few cases. Increases in C4ISR capability would be useful across the board, although there are variations in the type of C4I system required from future to future. The size of Army SOF either remained the same or increased in all of our cases. Significant investments in Army aviation help in all futures, but the mix between attack and lift helicopters varies considerably. The appropriate mix of deep strike capabilities also changes significantly from future to future; in some futures, we observed a need to orient deep strike systems toward engagements inside urban areas, where the technical requirements are quite distinct from those for in-depth engagements against enemy ground combat forces on an open battlefield.

ACKNOWLEDGMENTS

The authors want to express their appreciation to COL Matt Finley, Deputy Assistant Deputy Chief of Staff for Operations and Plans, Force Development. COL Finley reviewed earlier versions of the work and provided useful insights and direction. Additionally, we want to express thanks to David Kassing, the Director of RAND Arroyo's Strategy, Doctrine, and Resources Program, and to Richard Darilek, who co-led the project that sponsored this work. The authors would also like to thank technical reviewers Leland Joe of RAND and Stephen Kirin of MITRE for their constructive comments on earlier drafts of this documented briefing.

Alternative Futures and Their Implications for Army Modernization Planning

John Gordon IV and Brian Nichiporuk

April 2003

ARROYO CENTER RAND

This briefing draws upon research conducted for the Deputy Chief of Staff for Operations and Plans (DCSOPS) at Headquarters, Department of the Army. The research was part of a project on future force modernization carried out under the Strategy, Doctrine, and Resources Program of the Arroyo Center at RAND.

This presentation examines several alternative futures and the impact they could have on United States Army modernization planning.

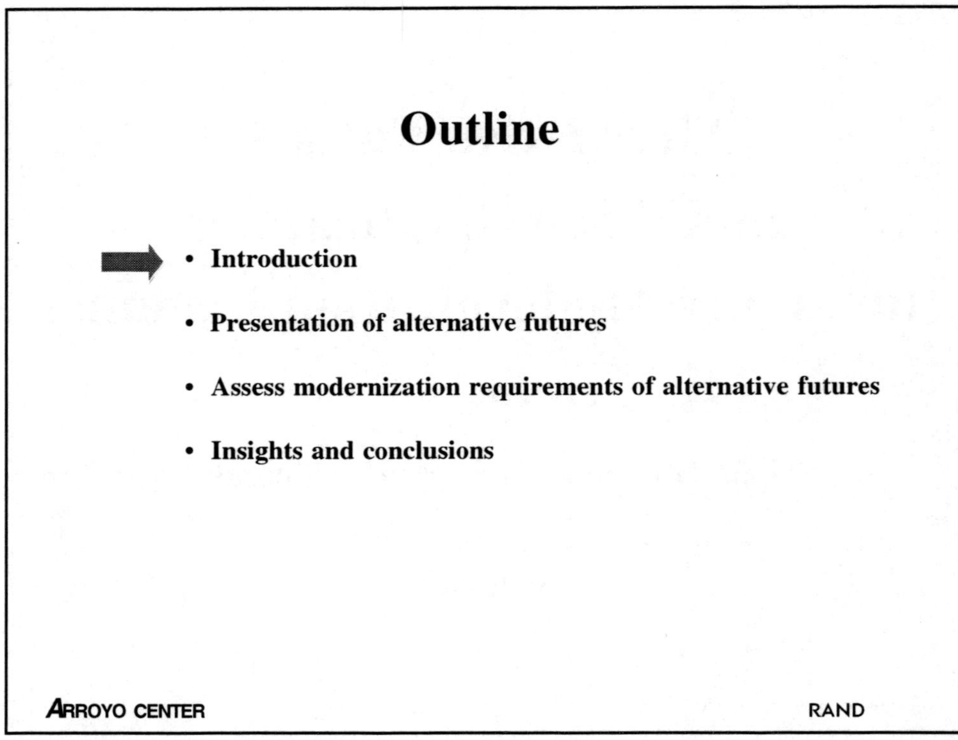

Our presentation unfolds in four parts. First, there is a brief introduction designed to explain the alternative futures paradigm. Second, we describe the six futures themselves and the main Army missions within each. Third, we assess the implications of each future for Army modernization investments and force structure/organization. Finally, we present some insights having to do with common modernization requirements across the spectrum of futures.

Motivation for Alternative Futures: U.S. Army Facing Major Decisions

- Post–Cold War drawdowncompleted
- Broad spectrum of operations probable in future; variations in intensity, scope, geography, duration likely
- Army Transformation Plan establishes marker for future; however, plan is not finalized yet
- Army's equipment stock is aging, importantrecapitalization decisions must be made in next decade
- Major equipment decisions have implications that last for decades

ARROYO CENTER RAND

There are a number of reasons why an examination of various alternative futures is important. Chief among them are the following:

The post–Cold War drawdown of U.S. military forces is generally seen as complete. America's armed forces are now focusing on different types of missions and threats. These forces are structured differently than they were during the decades of confrontation with the communist world.

A wide range of operations in the future is likely to be the norm. Missions will span the operational continuum from low-intensity peace operations to major theater wars (MTWs). Additionally, the geographic regions where U.S. military forces could operate will vary widely, as will the duration of the various operations that might be conducted.

The Army's Transformation Plan establishes a marker for the future. The details of the plan, however, are not yet finalized.

The Army's current equipment stock is aging. Most of the major combat vehicles and aircraft are purchases from the 1980s and early 1990s. As the Army's Transformation Plan makes clear, important decisions about recapitalization have to be made this decade.

Those decisions will have long-lasting effects on the Army. Given the post–World War II trends in military equipment, the systems that are placed in development and/or production during this decade will remain in the future Army for decades to come.

Use of Alternative Futures Can Help the Army Hedge Against Uncertainty

- **Trend projections are only useful for ten years or so, beyond that they have much less utility**
- **Use of a spectrum of futures enables one to see a range of possible modernization requirements for the Army**
- **Alternative futures makes it easier to pursue a hedging strategy**
  - **What common equipment needs might we see across a range of futures?**
  - **How appropriate is the emerging Transformation Plan for various possible futures?**

ARROYO CENTER RAND

The value of alternative futures as an approach to long-range planning is that it can help the Army hedge against uncertainty.

Various studies have indicated that trend projections are useful for only about ten years into the future. Beyond that they have far less utility, since there are so many factors (economic, political, technological, etc.) that could undergo significant change beyond ten years.

The use of alternative futures in long-range planning also enables an organization to foresee a range of possible modernization requirements. Different futures might require significantly different capabilities and, therefore, different modernization choices.

Importantly, the use of alternative futures makes it easier to pursue a hedging strategy as the farther future is considered. Planners can examine what common equipment needs are observed across a range of futures. Additionally, the Army could gain insights on how appropriate today's Transformation Plan is for various possible future strategic environments.

It should be noted that the six alternative futures used here have an artificial analytic clarity in that they each focus on one central feature of the international security environment (e.g., a near-peer competitor, transnationalism) as the organizing principle for Army force structure and modernization planning. The "real world" of today, by contrast, includes a complicated mix of threats that compete for planners' attention—terrorism, ethnic conflict, rogue states, etc. The authors recognize this divergence and thus present the alternative futures as distinct planning benchmarks for the 2025 timeframe that show the demands that each class of plausible future threat would place upon the Army.

There is, in our view, ample historical justification for an approach that focuses on a single, overarching threat type as the benchmark for American defense planning. After all, many of the international security environments that the United States has faced throughout its history presented one dominant concern that its defense establishment had to confront (e.g., Southern secessionism in the 1860s, conflicts with Native American tribes in the 1870s and 1880s, rival colonial powers in 1899–1914, Wilhelmine Germany in 1916–1918, British and Japanese naval expansionism in the 1920s, Nazi Germany in the 1940s, the USSR during the Cold War, rogue regional powers in the 1990s, Islamist terrorism in 2001–2002). We simply wished to give our worlds the same kind of focus as was enjoyed by military planners in some of those earlier periods.

If we had mixed classes of threat together, the linkages between each class and its impact upon Army force structure and modernization requirements would have become muddled and opaque to the research consumer.

Outline

- Introduction
- Presentation of alternative futures
- Assess modernization requirements of alternative futures
- Insights and conclusions

ARROYO CENTER RAND

Next we will examine the alternative futures that were considered in this study.

Alternative Futures
The Likely "What Ifs"

• U.S. Unipolarity • Democratic Peace	"Best" cases
• Major Competitor • Competitive Multipolarity	"Medium-Good" cases
• Transnational Web	"Medium-Bad" case
• Chaos/Anarchy	"Worst" case

ARROYO CENTER RAND

The six futures that we developed range from relatively benign cases—from the perspective of U.S. interests—to highly challenging strategic environments.

We have grouped these futures into four categories that include the following:

- The "best" cases are **U.S. Unipolarity** and **Democratic Peace**.
- The "medium-good" cases include the rise of a **Major Competitor** and **Competitive Multipolarity**.
- The **Transnational Web** represents what we consider a "medium-bad" case.
- Finally, the "worst" case is **Chaos/Anarchy**.

Each of these possible futures, and its possible implications for the Army, will be examined in this document.

Alternative Futures
Some Clarification

- Timeframe is 20–25 years in the future, not near term
- No assessment of the probability of each future—assumes that each possible future meets the "not implausible" test
- U.S. Unipolarity is closest to today's situation
- "Signposts" have been prepared for 2010 to help determine which futures are becoming more plausible
- Illustrative scenarios are available for each future

ARROYO CENTER RAND

Before examining each case in detail, it is helpful to clarify some of the assumptions underlying this approach.

The timeframe we are examining is roughly 20–25 years in the future. This is not a near-term trend projection.

There is no assessment of the probability of each future. The study assumes that each possible future could occur, and that each meets the "not implausible" test. That is, none of these futures represents a profoundly unlikely situation, such as a total breakdown of international society or collapse of modern civilization. Other than this very basic premise, there are no common macro-assumptions across the six futures.

The **U.S. Unipolarity** case is closest to today's situation. There is, however, no guarantee that this strategic situation will still be in force 20 or more years in the future.

Various "signposts" have been prepared to help determine which futures are becoming more plausible.[2]

Additionally, illustrative scenarios are available for each future.[3] The force size, structure, and mix discussions for each future that are presented in this document are based on some rough force-sizing analysis that was performed as part of a companion study that examined the implications of the six worlds for Army manpower and personnel policy.[4]

In terms of multinational issues, the force sizing done in these futures does explicitly assume that U.S. Army units are operating alongside allies and coalition partners. In none of the worlds and associated scenarios is a pure unilateralist military campaign posited. For example, in the first future (U.S. Unipolarity), both the operational concept and force package used by U.S. ground forces was predicated on substantial Malaysian military participation in the defense of Borneo against Indonesian attack.[5]

Also, we did not ignore the need to consider the contributions of the other services of the U.S. military in our alternative futures. We assumed that there would be no radical change in the Army's main roles and missions—in other words, we did not posit that either the Air Force or the Navy took any major roles or missions in the area of ground warfare away from the Army. However, we did assume that technological progress in the areas of tactical aircraft design, air-delivered PGMs, and UAVs will enable future regional conventional wars to be prosecuted successfully with smaller ground force packages than those that would have been necessary in the Desert Storm era. Thus, in those futures that emphasize traditional conventional warfighting on a large scale (like Major Competitor Rising), the Army force packages assigned are smaller than those that would have been deployed a decade ago. We simply believe that U.S. aerospace

---

[2]These signposts are not presented in this document, but are available in the related RAND work *Alternative Futures and Army Force Planning: Implications for the Objective Force Era*, Santa Monica, CA: RAND, MR-1676-A, 2003 (forthcoming).

[3]These are also available in *Alternative Futures and Army Force Planning: Implications for the Objective Force Era.*

[4]That work is presented in *Future Personnel Resource Management: Initial Report*, unpublished draft, RAND, August 1998.

[5]The details of the coalitions in which U.S. forces are assumed to operate in the different worlds are provided in *Alternative Futures and Army Force Planning: Implications for the Objective Force Era.*

power will improve enough to allow smaller Army forces to maneuver and mass fires more effectively against large, previous-generation threat ground forces.

Finally, it must be emphasized that our analysis here only examines implications of the six worlds for the active Army. Neither National Guard nor Army Reserve forces are covered. This is certainly a limitation in our study, but the decision was made to focus on the active force primarily because of time and resource constraints, and also because the Active Component would be affected earlier and more dramatically by any changes in the Army Transformation Plan than would the Reserve Components.

U.S. Unipolarity/Light Lethal Army

- **Features of this future**
  - **U.S. dominant militarily, economically, politically, culturally**
  - **China/Russia/India/EU do not challenge U.S. leadership**
  - **Regional powers hostile, intimidate neighbors with WMD and asymmetric strategies**
  - **Sporadic minor peace ops in areas of communal violence/famine**
- **Desired force characteristics**
  - **Land forces that can quickly deploy across large distances**
  - **Operational mobility with significant firepower, including deep fires, all-terrain combat vehicles**
  - **Significant deployable TMD and ISR capability**
- **Force size and composition**
  - **Total AC force size like today's**
  - **Army XXI units intended primarily for conventional combat against MTW-like opponents, new rapidly deployable units (Interim and Objective Force) for response to sudden crises**

ARROYO CENTER RAND

The first future is **U.S. Unipolarity**. We have described the associated U.S. Army force for this future as the *Light Lethal Army.*

The dominant features of this future include:

- The United States remains dominant militarily, economically, politically, and culturally.
- China, Russia, and India do not overtly challenge U.S. leadership, although they may be relatively hostile to U.S. strategic objectives.
- There are various regional powers that are hostile to the United States and its allies. Some of these regional powers are prepared to intimidate their neighbors with weapons of mass destruction and/or asymmetric strategies.
- The United States conducts sporadic peace operations in areas of "failed states" where there is famine and violence.

Given this environment, what would be the desired force characteristics of the U.S. Army? We envision that in this environment the Army would need capabilities that include:

- Forces that can quickly deploy over large distances to respond to various crises.
- Systems that are operationally and tactically agile in a wide variety of environments. Additionally, those systems will require high firepower, since the initially deploying force may be outnumbered in some circumstances.
- Significant, rapidly deployable, theater missile defense and intelligence, surveillance, and reconnaissance capabilities.

The force that we envision for this future is roughly the same size as today's Army. Remaining Army XXI–type units (upgraded versions of today's heavy armored and mechanized divisions and brigades) would be focused on conflicts at the level of major theater wars, while new organizations similar to those envisioned in the Army Transformation Plan (Interim and Objective Forces) would primarily be focused on response to sudden, unanticipated crises in locations where it is difficult to deploy the classic Army XXI–type organizations.

Democratic Peace/Policing Army

- **Features of this future**
  - **Spread of democracy eliminates risk of interstate wars**
  - **EU, China, Japan are major economic, but not political, rivals of U.S.**
  - **Residual social disorder and ethnic violence in parts of the Southern Hemisphere**
  - **Army participation in multinational stabilization ops**
- **Desired force characteristics**
  - **Protected light infantry, SF, airmobile units, MOUT, coalition interoperability**
  - **Robust urban intelligence gathering capability**
  - **Advanced nonlethal weaponry, personnel protection, and SIGINT**
- **Force size and composition**
  - **Total AC force size about two-thirds of today's**
  - **Post–Army XXI forces primarily infantry, MP, medical, SF, linguists, intel, air, and other support**

ARROYO CENTER RAND

Democratic Peace is a benign future. In this situation the efforts of the United States and its allies to spread democratic values have generally succeeded. There is little risk of major interstate warfare in this situation. The European Union, China, and Japan are economic competitors, but not political rivals of the United States.

There is residual social disorder and ethnic violence in some parts of the Southern Hemisphere that sometimes requires U.S. intervention. The U.S. Army, which we have titled the *Policing Army,* participates in various multinational stability operations in these areas.

The characteristics required of the Army in this future include:

- Protected light infantry forces, special operations units, organizations capable of operations in urban terrain, and the ability to regularly operate in coalitions.
- A robust intelligence-gathering capability in urban areas where many of the missions would probably be focused, since the world will be increasingly urbanized between now and 2025.

- Nonlethal weapons, individual soldier protection, and signals intelligence systems that would improve infantry capabilities in complex terrain.

The active Army of this future would be roughly two-thirds the size of today's force. The post–Army XXI units would be primarily high-tech light forces, military police–type organizations, special operations units, and associated support forces. All these units would be specifically trained for urban operations.

Major Competitor Rising/Big War Army

- **Features of this future**
  - **Single power or bloc rises to challenge U.S. across the board**
  - **Develops significant conventional and strategic nuclear capabilities including power projection and space; leverages RMA**
  - **Prepare to fight multi-corps-sized ground wars in two theaters simultaneously**
- **Desired force characteristics**
  - **Precision deep fires to defeat attacking echelons in concert with air power**
  - **Highly mobile force that can help defend areas threatened by peer-type opponent, and participate in counteroffensive operations**
  - **Advanced air/ground/space C4ISR architectures and attack helicopters**
  - **Ground-based theater ASAT and TMD capability**
- **Force size and composition**
  - **Total AC force size 10% larger than today's**
  - **Post–Army XXI forces intended for combat against peer-level opponent**

ARROYO CENTER RAND

A much more threatening future is **Major Competitor Rising**. This future would include a major power that is hostile to the United States and prepared to challenge it on a global basis. The opponent would probably invest heavily in power projection, strategic nuclear, and conventional land force capabilities. It is likely that this type of opponent would also adopt many of the technologies and operational concepts associated with what has become known as the Revolution in Military Affairs. This future could require that the United States be prepared to confront the opponent in major conflicts in more than one theater.

The Army of this future, which we term the *Big War Army,* would require the following kinds of characteristics:

- Precision deep strike to engage a large opponent throughout an operational area.
- The ability to defend areas of vital interest from a large, well-equipped, opposing land force, including the possibility of major counter-offensive operations to retake areas that might initially be lost in a war.

- Advanced command, control, communications, computer, intelligence, surveillance, and reconnaissance (C4ISR) capabilities. A powerful attack aviation capability.
- A theater missile defense system capable of countering large numbers of sophisticated enemy cruise and ballistic missiles.

The well-armed, high-tech Army of this future would be some 10 percent larger than today's Army. This is based on the assumption that the Army would be part of a joint U.S. approach to the major competitor in which there would be considerable reliance on Air Force- and Navy-delivered precision weapons.

Portions of the Army would probably be forward-deployed into areas of vital interest that are threatened by the major competitor. Those forward-deployed units and reinforcing forces would all be configured for combat against well-armed ground and air forces of this peer-level opponent.

Competitive Multipolarity/Global Maneuver Army

- Features of this future
  - 2–3 other great powers arise with military/economic capability comparable to U.S.
  - Defensive alliances, multisided competitions for influence in less-developed regions
  - Demand for power projection and presence deployments (Global Maneuver)
- Desired force characteristics
  - Diverse mix of combat capabilities for all terrain types
  - Ability to rapidly deploy force packages between and within theaters
  - Lean logistics and C2 networks that can support operations in all environments
- Force size and composition
  - Total AC force size about 50% larger than today's
  - Post–Army XXI forces designed for rapid deployment to threatened regions; Army XXI–type forces mostly used where prepositioning is feasible and as counteroffensive reserve

ARROYO CENTER RAND

Another threatening future would be the **Competitive Multipolarity** world. Here the Cold War–era alliances have broken up and two or three new power blocs have emerged. NATO is gone, and some of the U.S.'s old European allies are now aligned with a resurgent Russia. In Asia there is another power bloc with China as its hub. Both of these competitive alliances have roughly the same power as the United States and its allies. Given that conflict could break out in many different parts of the world against either of these competitive power blocs, the United States would require forces capable of global maneuver.

The characteristics required of this *Global Maneuver Army* would include:

- Ability to conduct operations in many different climates and terrain types.
- Ability to rapidly deploy forces to and between different theaters of operations.
- Relatively lean logistics requirements and C4ISR systems that can support operations in all environments.

The active Army of this future would probably be roughly 50 percent larger than the 2000 force. The post–Army XXI forces would be designed for rapid deployment to various locations. Remaining Army XXI–type units would be located in those areas of vital interest where prepositioning is feasible, as well as serving as a counteroffensive force.

Transnational WebNetwar Army

- **Features of this future**
  - **At expense of nation-states, rise in power of transnational distributed actors (e.g., corporations, criminal syndicates, terrorist or special interest groups, ethnic diasporas)**
- **Desired force characteristics**
  - **Robust cyberintelligence/IW capability**
  - **Well-funded "crisis public affairs" branch**
  - **Networked teams of immediately deployable SOF personnel**
- **Force size and composition**
  - **Total AC force size nearly 40% smaller than today's**
  - **Post–Army XXI forces primarily Rangers, SF, computer/electronics experts, and intel, large portion of the Army not intended for conventional combat operations**

ARROYO CENTER RAND

A very different future is **Transnational Web**. Here the role of the nation-state has withered to a large extent. Transnational organizations, some with criminal and/or terrorist intent, have risen in power. Some of these transnational groups represent ethnic or other groups whose interests span the borders of many nations. These groups are capable of various serious asymmetric actions against governments, be they powerful or weak states.

The type of Army required in this future is quite different from a conventional force oriented toward fighting the army of another nation. Essential characteristics would include:

- A very robust cyberwar and electronic intelligence capability.
- A "crisis public affairs" branch to participate in information operations.
- Teams of special operations forces capable of operating against nonstate actors who frequently employ electronic means to wreak damage on state and nonstate opponents.

This *Netwar Army* would be considerably different from today's force, possibly 40 percent smaller. The post–Army XXI elements would be mostly Rangers, Special Forces, and electronics and computer experts. Indeed, large portions of this Army would not be intended for conventional combat against the armed forces of another nation.

Chaos-Anarchy/Dirty Environment Army

- **Features of this future**
  - **In several regions, nation-states destroyed by overpopulation, environmental degradation, ethnic violence, emergence of regional warlords**
  - **Increase in terrorism, WMD proliferation, mass migrations, disease**
- **Desired force characteristics**
  - **Force protection and counterterrorismcapabilities against NBC use**
  - **Emphasis on light infantry with light armored vehicles and heliborne support**
  - **Strong infrastructure restoration and MOUT capability**
- **Force size and composition**
  - **Total AC force size about 10% smaller than today's**
  - **Post–Army XXI forces primarily infantry, Rangers, engineers, medical, MP, SF, and other support (including linguists, intel, and air)**

ARROYO CENTER RAND

The final future we examined was termed **Chaos-Anarchy**. Although there is no peer-type opponent in this future, in many respects this future is the most bleak. It postulates many failed states and widespread problems in much of what we today call the Third World. Overpopulation, environmental degradation, and ethnic problems are endemic in many parts of the world. These problems have led to many regional conflicts, civil wars, and mass migrations. The presence of weapons of mass destruction and spreading infectious diseases has made the situation particularly serious in some regions.

The *Dirty Environment Army* of this future would require these kinds of characteristics:

- Force protection and counterterrorism capabilities, including counter-WMD abilities.
- Light forces, with associated combat vehicles and air support, that can operate in urban situations.

- Infrastructure-restoration capabilities for use in areas where the collapse or destruction of the local infrastructure has contributed to human suffering and conflict.

The active Army of this future would probably be some 10 percent smaller than today's force. The post–Army XXI elements would be focused on infantry, support engineers, special operations, and various support units that would be capable of operations in far-flung areas of the globe.

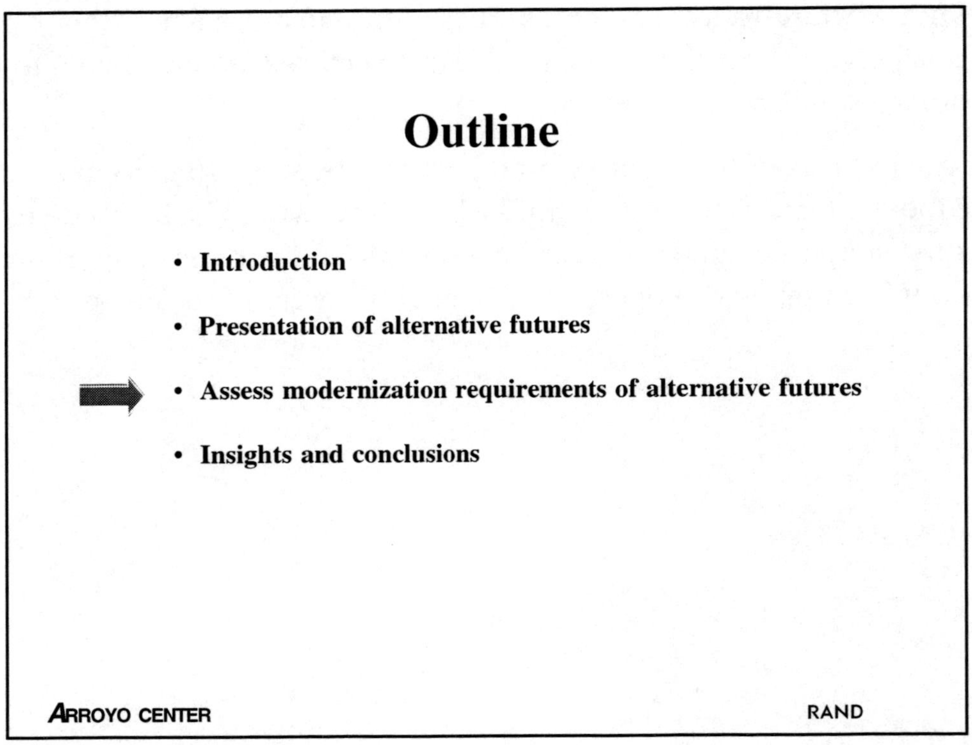

Now that the various futures have been reviewed, we will examine their individual modernization requirements.

Assessing Modernization Requirements for Alternative Futures

- Examine types of forces and key capabilities
  - Army XXI, Interim, Objective, SOF
  - TMD, C4ISR, deep strike
- Determine the magnitude of change each future could impose on the current Army Transformation Plan
- Assess whether generally the same, more, or less resources would be required for each future

Same level of resources
Modest increase in resources
Modest decrease in resources
Significant increase in resources
Significant decrease in resources

- Modest increase or decrease implies 5–10% change from today's level of resources
- Significant increase or decrease is greater than 10% change

ARROYO CENTER RAND

We used a qualitative approach to assess the modernization implications of each future. Certain Army forces and capabilities were examined in an attempt to determine whether more or less would be needed in each future. Specifically, we examined the implications of each future for modernization spending in the following categories: Army XXI forces, Interim forces, Objective-level forces, SOF, theater missile defense (TMD), aviation, combat service support (CSS), WMD defense, C4ISR, and deep strike. The first eight are self-explanatory; however, the last two require definition. C4ISR in this document encompasses a wide range of sensor, communications, and battlefield information-processing technologies. Although this category is broad, it is also regarded as distinct by many military analysts because it describes technologies that are currently driving the Revolution in Military Affairs. These are the technologies that have to do with situational awareness throughout a deployed force. The authors believe that any disaggregation of this category would make it more difficult for readers to compare the importance of situational awareness to that of firepower, mobility, support, and protection in future visions of the Army. Deep strike capabilities, as defined here, are ground-based, long-range, precision-strike capabilities like MLRS, ATACMS, BAT, and Copperhead. Our deep strike category does not include long-range attack aviation.

In determining the magnitude of change that could result, we elected to label a level of change as either modest or significant. A modest change means that no more than a 5–10 percent change in structure or resources would be needed in that area to accommodate the needs of a particular future. A significant change was more than 10 percent change in that category.

Based on what is known about the current Army Transformation Plan, and the forces that it would provide by roughly 2020, we attempted to determine whether a modest or significant change would be warranted. In some cases it appeared that no change would be required in resources or structure, but there might be a need to refocus resources. For example, in some of the futures it appeared that there would be a major role for Army aviation, but it might require a bias toward lift-type aircraft as opposed to attack aircraft. Those nuances will be highlighted in the text.

Finally, in the following section we describe in detail the force organization we propose for the Army in each of the six futures. Our organizational concepts are based on numbers of divisions. We have chosen the division as our metric because it is still the foundation unit that the Army senior leadership uses in public discussions of service strength and resource management today vice the Cold War era. The division is also widely used as a reference measure by civilian defense analysts and scholars. We are aware that, in reality, the Army includes other unit types (e.g., SF Groups, SBCTs, etc.) but have consciously chosen to use the division as the "currency of the realm" in this document because it makes our analysis transparent to a large audience of analysts both inside and outside the Department of Defense.

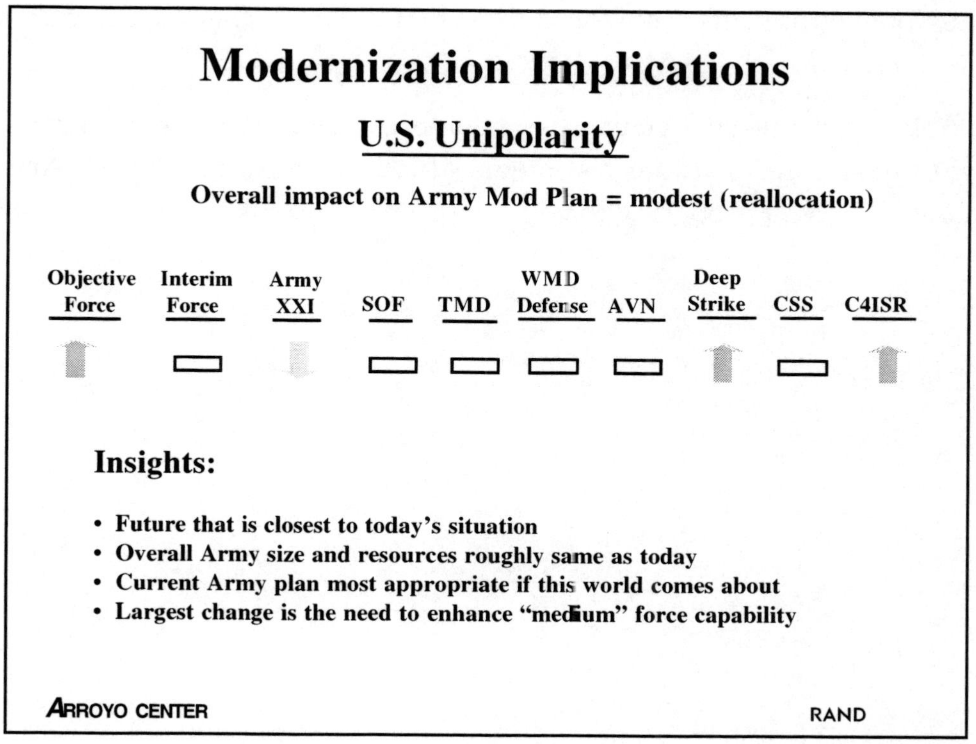

This chart shows our assessment of the implications of the **U.S. Unipolarity** future. The categories shown on this chart remain constant in the next five futures.

The essential mission of the Army in this future would be to rapidly deploy for crisis-response missions in various locations around the world. Opponents would be either of the current MTW type (e.g., Iraq) or smaller. There would be no peer-level opponent in this future.

We saw a need to modestly increase the Objective Force in this situation. This should result in a capability to rapidly defeat regional opponents. Army XXI units would be reduced by a modest amount. In overall terms, this means that by roughly 2020 one additional division would be converted to Objective Force configuration at the expense of older Army XXI structure.

Most other categories appeared to require little if any change. We determined that modest increases in deep strike (e.g., additional ATACMS) and C4ISR would be warranted in this situation, since increases in both those areas would contribute to achieving overmatch of the regional opponents that would represent the most challenging case in this future.

That relatively little change was deemed necessary in this case is because this future is closest to today's strategic situation—the situation the Army is in the process of preparing for.

Illustrative Army Structure
U.S. Unipolarity
Major Active Army Maneuver Units

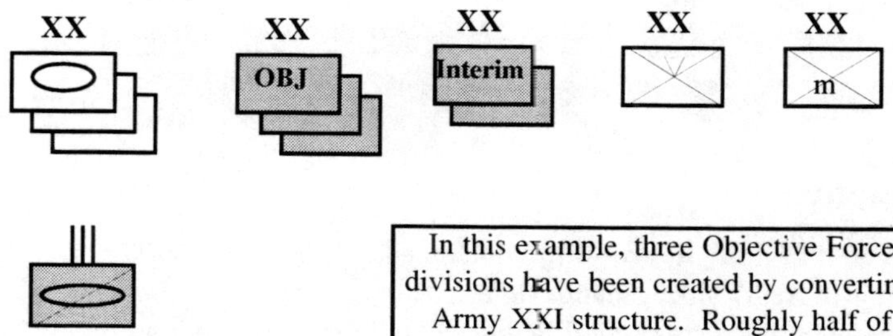

In this example, three Objective Force divisions have been created by converting Army XXI structure. Roughly half of structure is "post–Army XXI"

Note: Shading indicates "post–Army XXI" (Objective and Interim Forces); white indicates that the unit is at generally Army XXI levels of capability.

ARROYO CENTER RAND

This chart illustrates the key elements of active Army structure in the **U.S. Unipolarity** future. The structure shown above represents what the Army of this future might look like in the 2020–2025 period.

The Transformation organizations (Interim and Objective Force) are shaded, while the older Army XXI–type units are in white.

The Army is roughly the same size as today. The post–Army XXI units would focus on rapid deployment to respond to crisis situations, operating in conjunction with air assault and airborne forces. Army XXI forces would be available for forward presence missions and for large-scale counteroffensive missions in situations where Interim and/or Objective Forces are not adequate.

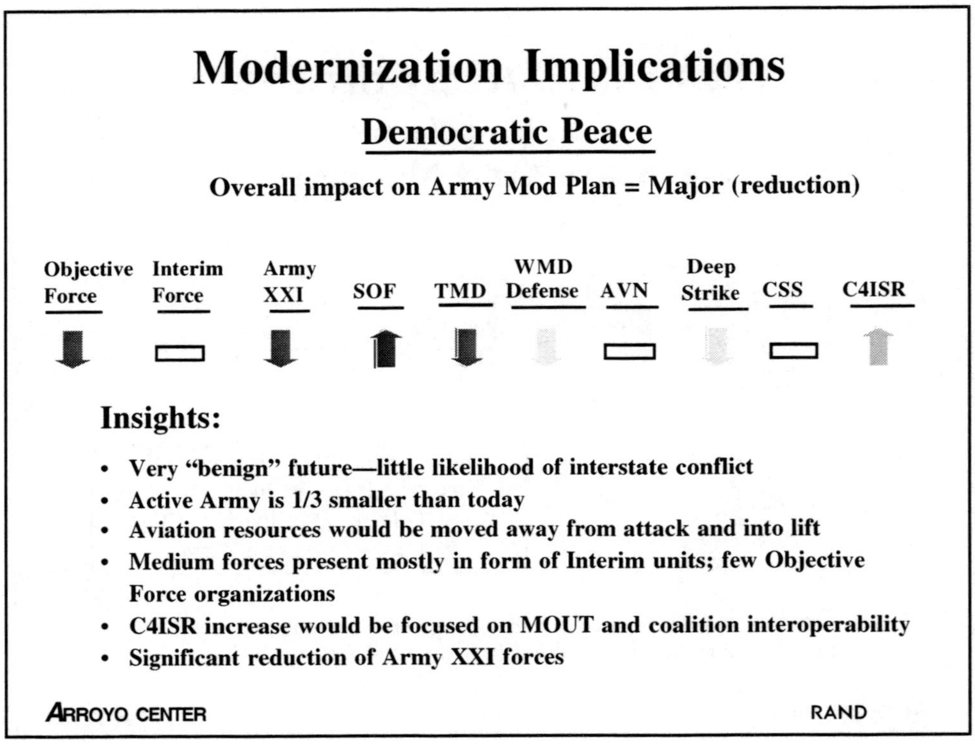

The Army of the **Democratic Peace** future differs significantly from today's force.

Since the world is a generally peaceful place in this future, the Army is not only smaller, it is focused on different types of missions.

Due to the absence of major regional threats, we determined that there would be significant reductions in traditional Army XXI forces, as well as a major truncation of the potentially expensive Objective Force organizations that the Army is currently exploring. Additionally, TMD would be reduced significantly in this future due to the absence of regional powers with large numbers of missiles. Similarly, there could be modest reductions in counter-WMD capabilities and deep strike assets.

The Interim forces that the Army is currently creating would be very appropriate for this future, since threats would be relatively low. We also determined that there would be a need to increase C4ISR capabilities, since Army forces might have to rapidly deploy to an unforeseen humanitarian crisis in an unfamiliar environment.

In this future the Army should maintain the current suite of combat service support capabilities in order to perform peace operations and humanitarian missions. This is a case where resources for Army aviation would remain roughly the same as today, but with an important difference. Here we saw a need to refocus Army aviation from attack capabilities to lift. Such a change would help enable low-intensity missions by Army forces.

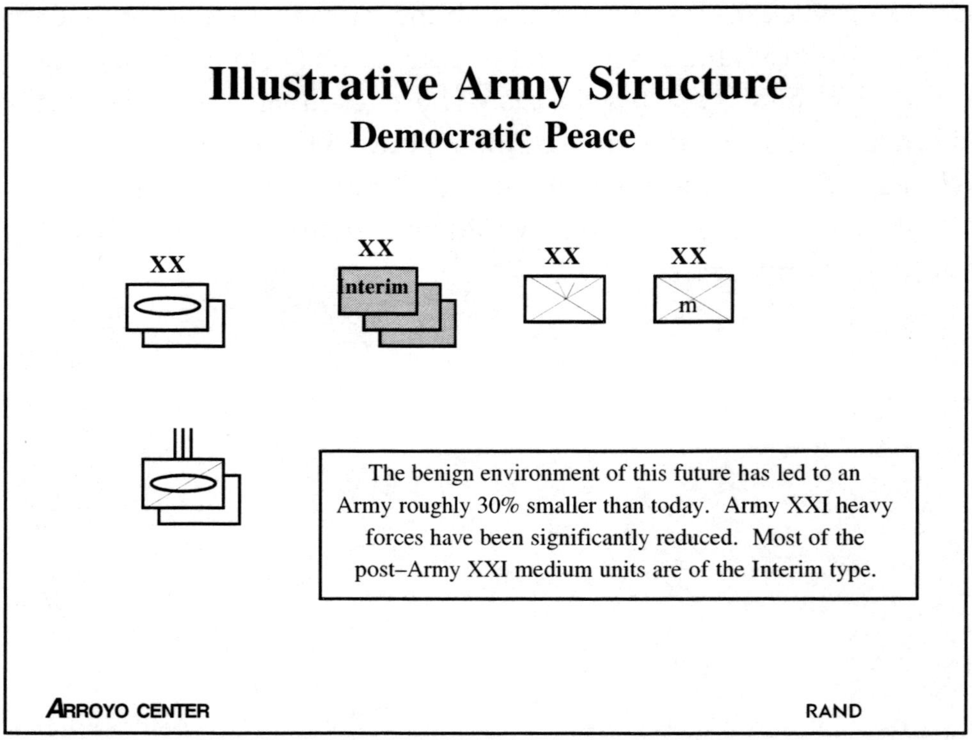

This chart illustrates the type of Army that would be appropriate for **Democratic Peace**. The post–Army XXI forces are of the Interim type. We saw relatively little need for Objective Force organizations in this future. A small number of Army XXI heavy forces are retained as a hedge against the possibility of a war with a regional power—a remote event in this future.

The active force is some 30 percent smaller than today's Army. Most of the modernization effort has been devoted to the creation of several Interim divisions.

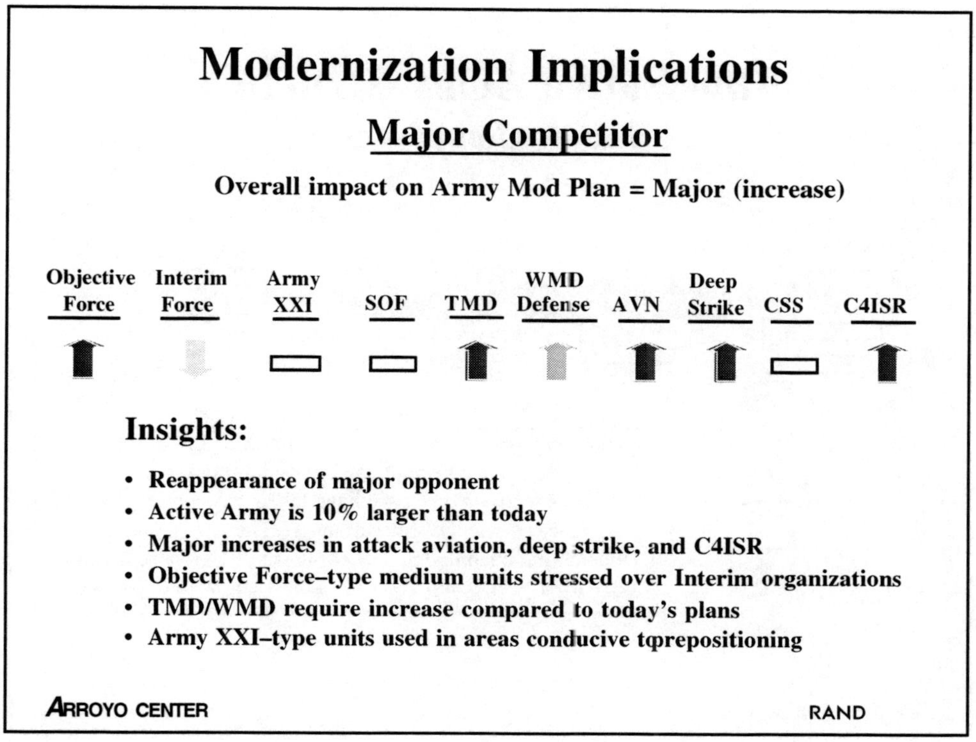

Major Competitor is a threatening future. This situation would require the Army to prepare to deter and, if necessary, defeat a peer-level opponent.

We envisioned major increases in Objective Force, TMD, aviation (in this case attack), deep strike, and C4ISR capabilities. Conversely, the less-capable Interim units would decrease in this future.

Army XXI forces would be employed in areas where they could be prepositioned for deterrence purposes. We envisioned that SOF and CSS capabilities would remain roughly the same in this situation.

Since areas of confrontation with the peer opponent could be determined in advance, the Army could plan for operations in those locations, as was the case in the Cold War. Due to the potential power of the major competitor's ground and air forces, it would be essential that Objective Force units come equipped with the systems and attributes necessary to defeat this class of opponent.

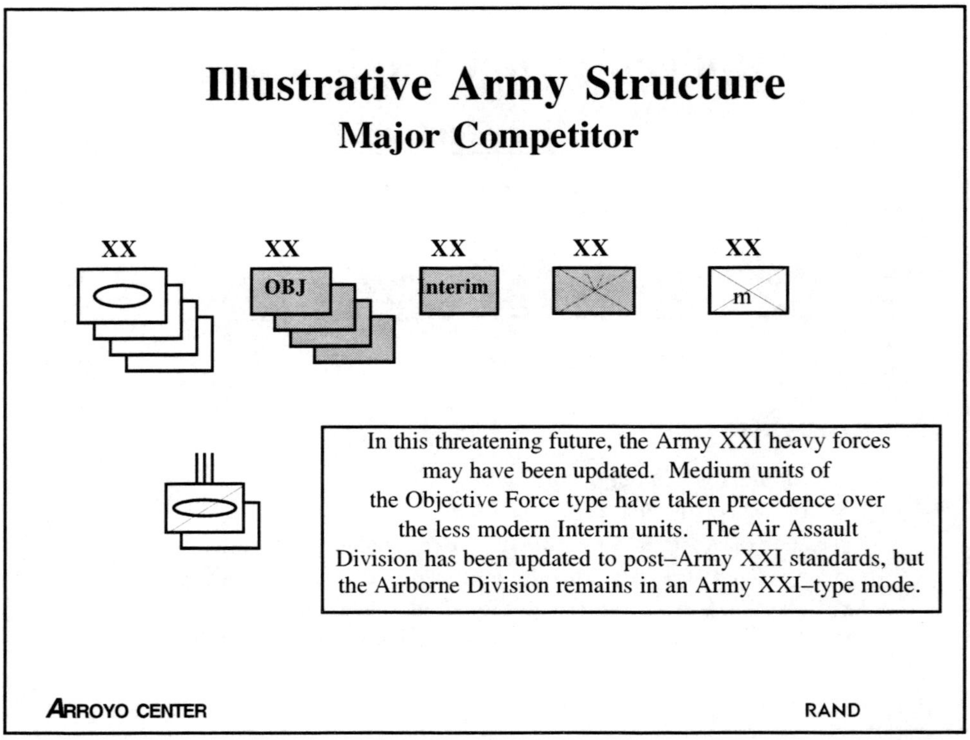

The Army in the **Major Competitor** case could look like the one above. This force is roughly 10 percent larger than today's Army, but it has been modernized to Objective Force standards to a considerable extent.

In this situation, where the geographic areas of potential confrontation with the major competitor could be identified, Army XXI units could be positioned for deterrence purposes. Additionally, we envision that these Army XXI units would be modernized to a greater extent than is envisioned today.

There are a minimal number of Interim forces in this structure, the more capable Objective Force units having received priority.

We also envisioned that the air assault division would be upgraded to post–Army XXI standards with new attack and lift aviation systems. Meanwhile, the airborne division would probably be intended for crisis-response missions in areas outside the major competitor's sphere of influence, and could therefore remain at roughly Army XXI levels of modernization.

Modernization Implications

Competitive Multipolarity

Overall impact on Army Mod Plan = Major (significant increase)

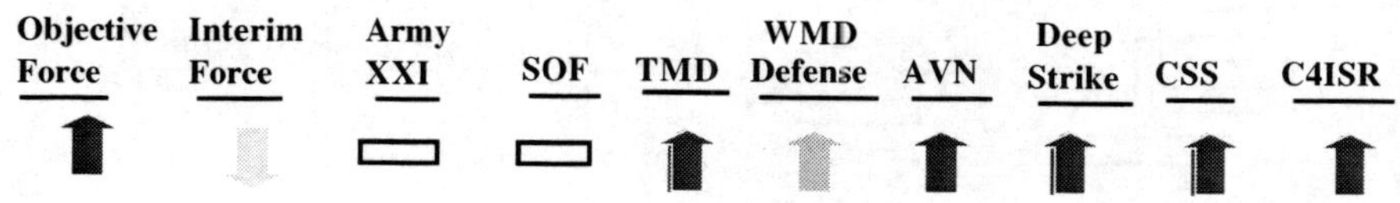

Insights:

- **U.S. faced with two or three potentially hostile power groups**
- **Active Army is 50% larger than today**
- **Deep strike, medium forces, CSS, and C4ISR receive significant increase in resources**
- **Additional CSS capabilities needed due to large number of possible deployments**
- **Medium units introduced in quantity**
- **TMD/WMD require some increase over today's plans**
- **Army XXI forces used in areas where prepo feasible and for counteroffensive**

ARROYO CENTER RAND

The **Competitive Multipolarity** future could lead to a significantly larger Army. In this future the Army would have to be prepared for deployments to confront hostile actions by two different coalitions that are opposed to the United States. Operations in many parts of the world would be the norm in this future.

Due to the need for rapid-deployment capability and the requirement to combat capable units of the opposing coalitions, we envisioned a significant increase in Objective Force units. There would also be a need to make major increases in TMD, aviation (lift and attack), deep strike, and C4ISR capabilities.

Increases in counter-WMD and CSS capabilities would also appear to be warranted in this situation. Certainly the opposing coalitions would have the option of employing WMD against the United States, and the need to operate in many different regions would require enhanced CSS capabilities in the active force.

We envisioned a modest reduction in Interim units in this situation, with more powerful Objective Force units being favored.

SOF and Army XXI forces were assessed to be appropriate at today's levels.

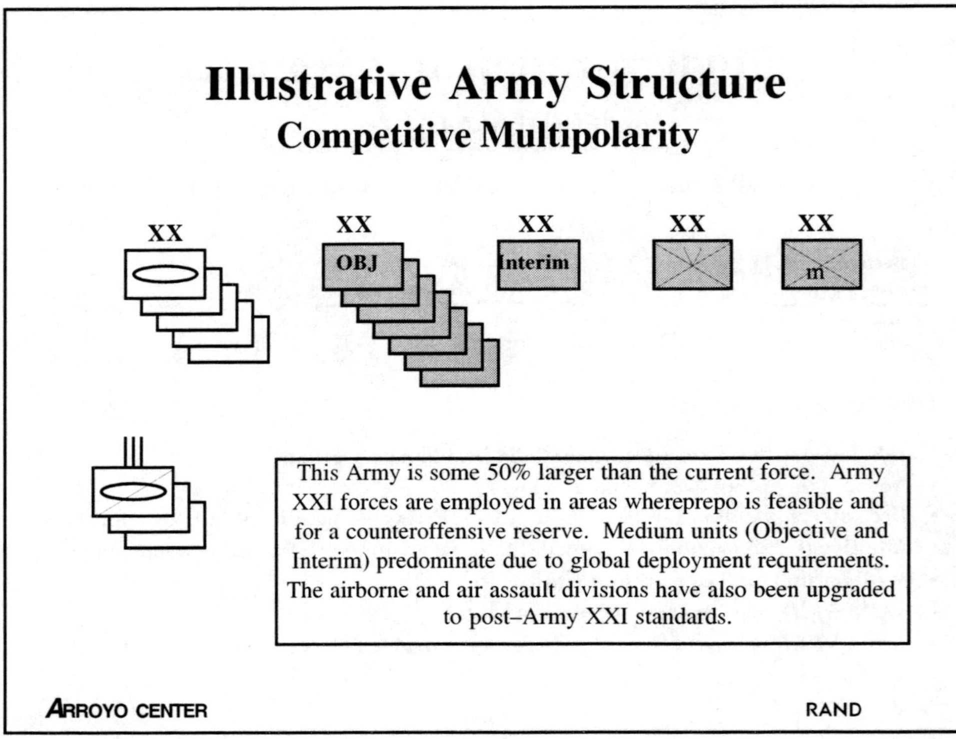

This Army is considerably larger than today's force, roughly 50 percent larger in our estimate. "Medium" units (primarily Objective) predominate in this threatening future.

The traditional Army XXI heavy forces that are available are either prepositioned in areas of vital interest as a deterrent or are held as a strategic counterattack reserve.

In this situation the airborne and air assault divisions have also been upgraded to post–Army XXI standards due to the need to rapidly deploy into areas where confrontation with the opposing coalitions is possible. These two units could be raised to either Interim or Objective Force standards, depending on resources and the role that is envisioned for them in this strategic environment.

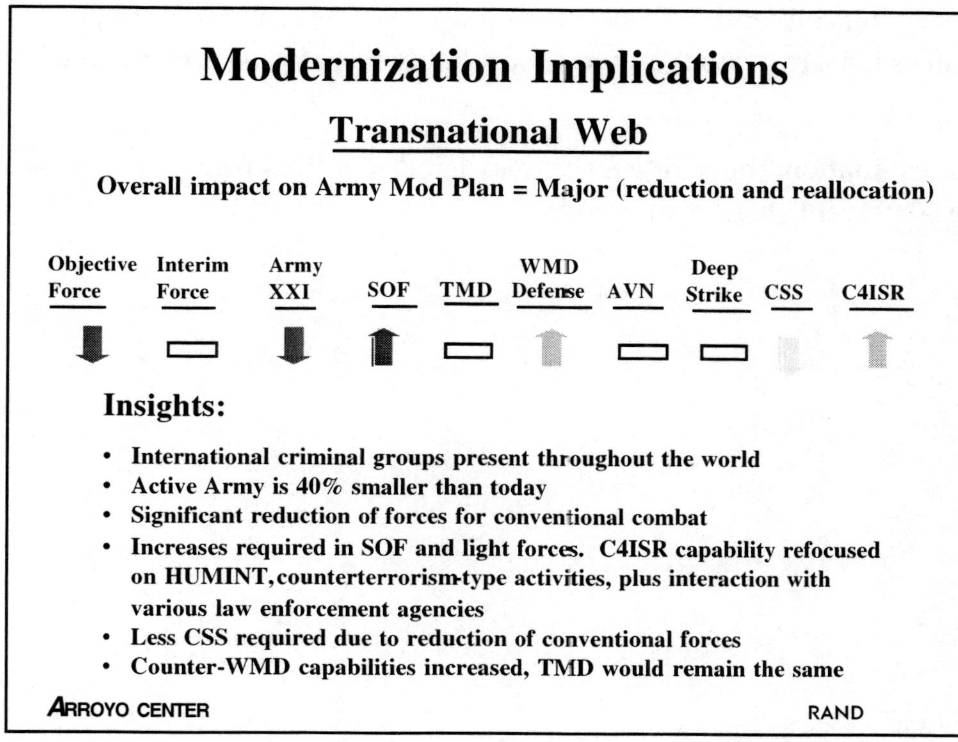

Transnational Web is a future where the threats to U.S. interests differ fundamentally from those we plan for today. The chances for major interstate warfare are relatively low. Rather, major portions of the U.S. armed forces have been reoriented to respond to the challenges posed by hostile transnational entities such as terrorist groups.

Given this strategic situation, we determined that there would be far less need for Objective Force and Army XXI–type forces. Medium units of the Interim type would probably be adequate for the intervention missions that might be required in this future.

On the other hand, SOF capabilities would require a very significant increase over today's levels. These SOF units of the future would be focused on countering the actions of various nonstate organizations. Counter-WMD and C4ISR capabilities would also require modest increases in this future.

We determined that TMD, aviation, and deep strike could remain at generally the levels of today, although there would be a need for some reprioritization. This could include refocusing deep strike on missions to engage hostile groups that have been located inside urban areas rather than striking conventional enemy land forces at depth. Similarly, resources for Army aviation would probably be refocused from attack to lift.

In our estimation, the active force associated with this future would be some 40 percent smaller than today's.

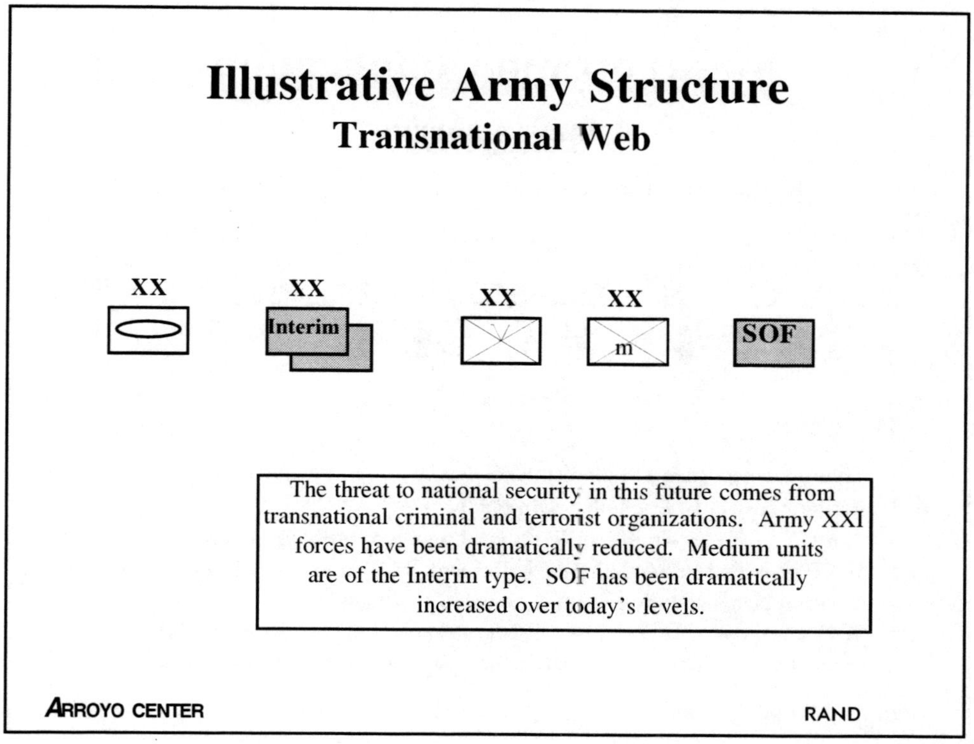

The conventional combat component of this Army is significantly smaller than today's force. The post–Army XXI elements are represented by the Interim units. A very small Army XXI heavy component is retained as a hedge; most heavy capability would reside in the Reserve Components in this future.

Although the details are not apparent on this chart, we envisioned that a very significant increase in special operations forces would be required in this future. Those SOF elements would require specialized capabilities to cope with various types of threats posed by transnational entities.

Modernization Implications

Chaos-Anarchy

Overall impact on Army Mod Plan = Moderate (reallocation)

Objective Force	Interim Force	Army XXI	SOF	TMD	WMD Defense	AVN	Deep Strike	CSS	C4ISR

Insights:

- Chaotic world, many "failed states"
- Active Army is 10% smaller than today
- Army XXI forces significantly reduced, medium units predominate
- Much greater counter-WMD requirement, TMD may be reduced slightly
- Aviation refocused toward lift as opposed to attack
- SOF, CSS, and C4ISR all increased, probably with focus toward MOUT
- Deep strike reduced somewhat, refocused toward MOUT requirements

ARROYO CENTER RAND

The **Chaos-Anarchy** future probably represents the worst case for the world as a whole. In this future there are many failed states, and much conflict in what we today call the Third World.

In this case, we determined that there would be a need for significant increases in SOF, counter-WMD, and CSS capabilities in order to provide the capabilities for interventions in various regional crises. Many of these interventions would be for humanitarian purposes.

C4ISR capabilities would probably merit a modest increase, as well as a refocusing toward providing situational awareness in chaotic failed-state environments.

Objective and Interim Forces would merit a modest increase in this situation, as would Army aviation. As was the case in some previous futures, this is also a situation where aviation resources might have to be refocused from attack toward lift.

TMD and deep strike could be reduced in this case, since the areas where intervention would be required would probably have weak military forces.

We saw the largest reduction in this future being in the area of Army XXI heavy forces. With no major competitor, and the focus of Army operations being rapid interventions into failed states, there would be relatively little need for traditional Army XXI heavy units in the active component.

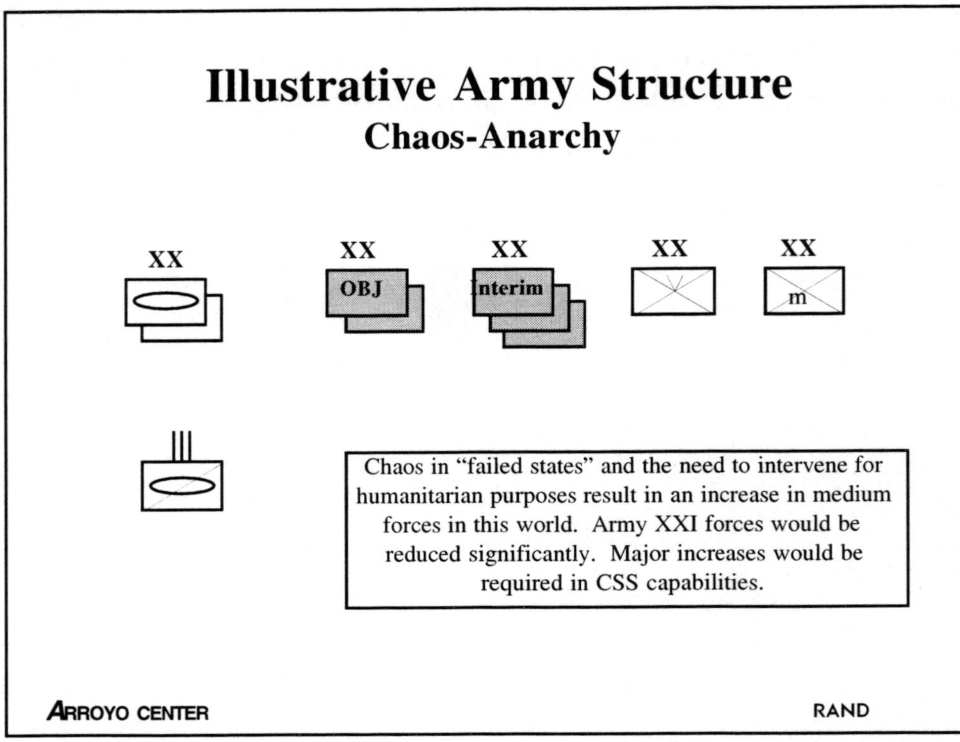

This Army is some 10 percent smaller than today's force. It has upgraded several of its divisions to post–Army XXI standards.

Medium forces are favored in this future due to their ability to rapidly deploy and operate in areas where the infrastructure is relatively poor. The airborne and air assault organizations have been retained for intervention missions, but have been held at Army XXI levels of modernization.

The few remaining Army XXI heavy forces are envisioned as a hedge against the remote possibility that Interim or Objective units might be inadequate in some circumstances.

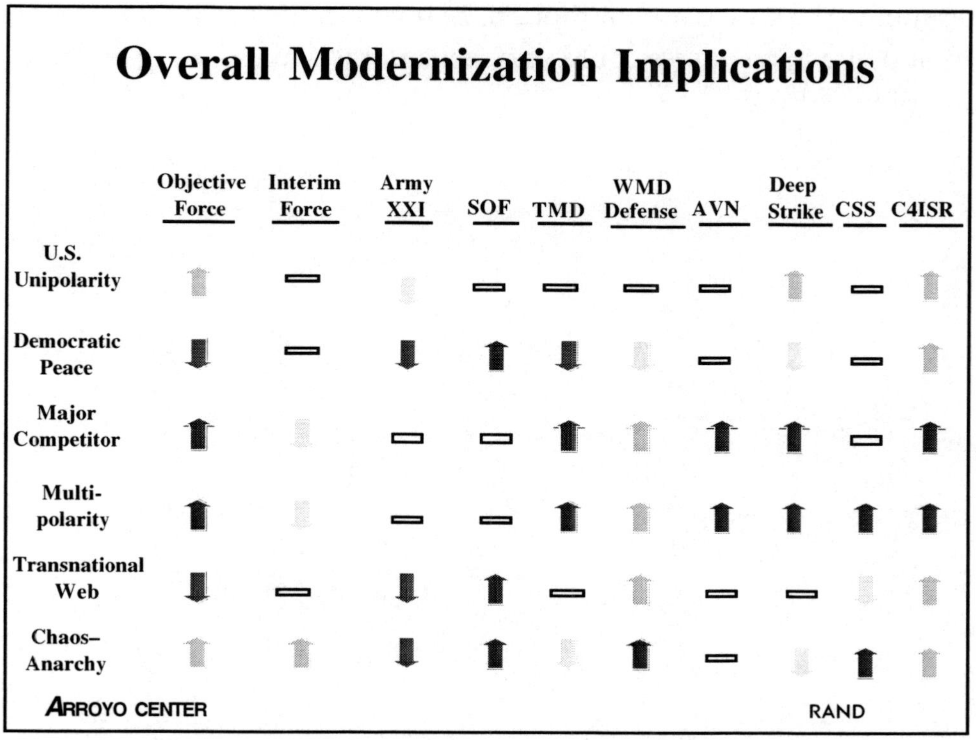

This chart summarizes the six futures we examined.

The first case, **U.S. Unipolarity**, differs little from today's plan, since that future is closest to today's strategic situation.

All the other futures show a greater degree of divergence from today's plans. It is interesting to note that in our estimation, C4ISR merited some amount of increase in all cases, although different types of C4I would be required in the various futures. Similarly, SOF either stayed the same or was increased in all cases.

Traditional Army XXI forces (particularly heavy forces) either remained the same or declined, depending on the future. In no case did we determine a need to increase this force type.

Objective Forces went from major decreases in the futures where there was little possibility of combat against conventional enemy armed forces (**Democratic Peace** and **Transnational Web**) to significant increases in the **Major Competitor** and **Competitive Multipolarity** cases.

TMD and WMD capabilities varied depending on the future, as did deep strike. Army aviation capabilities either remained the same or were recommended for increases, although, as noted in the earlier text, the mix of lift and attack resources could vary considerably depending on the nature of the future.

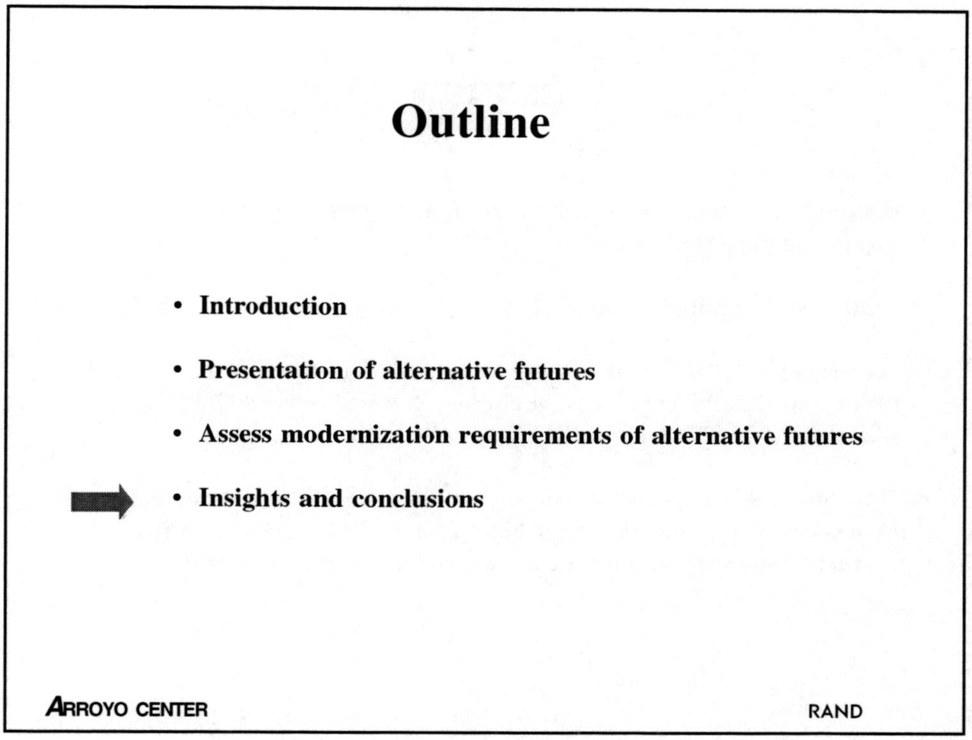

The final section of this briefing provides our overall insights and conclusions.

Insights

- Medium forces (Objective and Interim), are appropriate in a majority of possible futures.
- Continued investment in Army XXI–type forces is appropriate in few cases.
- Increases in C4ISR capability are appropriate in all the futures, albeit in some worlds the capability would be refocused toward OOTW-type situations. (OOTW = operations other than war.)
- TMD and counter-WMD improvements vary somewhat. In some futures there could be a requirement for high level of WMD protection from "terrorist" sources, without a commensurate increase in missile defenses.

ARROYO CENTER RAND

From our examination of six different alternative futures it appeared that medium forces (Objective and Interim) are appropriate in a majority of possible futures. In some cases they appeared to offer enhanced combat capabilities (particularly the Objective Force) against threatening opponents, while in other cases their deployability and ability to operate in areas with poor infrastructure were desirable capabilities.

From our assessment of the future cases, continued investment in Army XXI capabilities was appropriate in only a few situations.

Increases in C4ISR capability seemed appropriate in all cases, although there would be important differences as to the type of C4I system required. For example, a C4I system optimized against a major conventional opponent (e.g., the **Major Competitor** case) might not be at all appropriate in the **Transnational Web** situation.

TMD and counter-WMD capabilities vary somewhat. In some futures there would be a requirement for a high level of WMD protection, but little need for a missile defense capability (e.g., **Transnational Web**).

Insights
continued

- **SOF remains the same or warrants an increase in all cases.**
- **Significant investment in Army aviation is appropriate in all futures, but in several futures the aviation mix would change in favor of lift as opposed to attack aircraft.**
- **The mix of deep strike capabilities could change considerably in various futures. In futures with little possibility of engaging enemy armies the need for anti-armorPGMs would decline dramatically. MOUT-focused deep strike capabilities may be in higher demand in some futures.**

ARROYO CENTER RAND

Special operations forces capabilities remained at present levels or required increases in all cases. Indeed, in the case of the **Transnational Web** future, the need for increased SOF-type capabilities would be paramount.

A significant investment in Army aviation appears appropriate in all futures, although the investment mix could vary considerably between attack and lift assets.

Similarly, the mix of deep strike capabilities could also vary considerably. In some of the futures there might be a need to refocus deep strike systems toward engagements inside urban areas where building penetration and the potential for minimal collateral damage would be of paramount importance, rather than engaging enemy ground combat forces at depth on the battlefield.

Conclusions

- **The future is uncertain—beyond 10 years, trend projections become difficult.**
- **The Army's capital stock is aging; major decisions will be required in the next decade.**
- **Major systems tend to stay in the force for decades—today's decisions will impact the 2030s or later.**
- **Long-term Army modernization strategy should accommodate the possibility of very different futures—systems and platforms with multiple uses may be the wisest long-term strategy.**

ARROYO CENTER RAND

In conclusion we would offer the following observations:

- The far future is uncertain. Beyond roughly 10 years, trend projection becomes problematic. It is very difficult to determine what the strategic landscape will look like 20–25 years in the future.
- The Army's capital stock—its fighting vehicles and aircraft—is aging. During the current decade, major decisions will have to be made on replacement systems. The Army Transformation Plan will have to be finalized and many of the associated decisions made.
- Major systems tend to remain in service for decades. The systems that are put into production and development in this decade will be in the Army of 2030 or later.
- Long-term Army modernization strategy should accommodate the possibility of various futures in the next 20–30 years. Systems and platforms that have applicability in many different plausible futures may represent the wisest investment strategy.

Conclusions
continued

- C4ISR enhancements, medium forces, SOF, and a mix of aviation capabilities appear to be sound long-term investments.
- Heavy force modernization (Army XXI upgrades) seems inappropriate for several of the possible futures.
- Need to systematically revisit the signposts and indicators that define the various futures.

ARROYO CENTER RAND

- C4ISR enhancements, medium units (Interim and Objective), SOF, and a mix of aviation capabilities appear to be sound long-term investments across a range of possible futures.
- Army XXI enhancements, specifically efforts to upgrade the heavy component, appear to be inappropriate in several of the futures.
- There is a need to systematically revisit signposts associated with the futures employed in this analysis in order to determine, as time progresses, which direction the world of 2025 actually appears to be taking.

BIBLIOGRAPHY

Bevin, Alexander, *The Future of Warfare,* New York: W.W. Norton, 1995.

Engelbrecht, Joseph A., et al., *Alternate Futures for 2025: Security Planning to Avoid Surprise,* Maxwell Air Force Base, AL: Air University Press, 1996.

Kendall, Frank, and Paul Funk, *Full-Spectrum Projection for 2025-Era Ground Combat Vehicles,* a final report, briefing charts, FY99 Army Science Board Summer Study, July 1999.

Kugler, Richard L., *Changes Ahead: Future Directions for the U.S. Overseas Military Presence,* Santa Monica, CA: RAND, MR-956-AF, 1998.

Strategic Assessment 1998: Engaging Power for Peace, Washington, D.C.: National Defense University, 1998.